Charles Barney Cory

A Naturalist in the Magdalen Islands

Charles Barney Cory

A Naturalist in the Magdalen Islands

ISBN/EAN: 9783744716819

Printed in Europe, USA, Canada, Australia, Japan

Cover: Foto ©berggeist007 / pixelio.de

More available books at **www.hansebooks.com**

BYRON ISLAND.

MAGDALEN ISLANDS;

GIVING

A DESCRIPTION OF THE ISLANDS AND LIST OF THE BIRDS TAKEN
THERE, WITH OTHER ORNITHOLOGICAL NOTES.

BY

CHARLES B. CORY.

ILLUSTRATED FROM SKETCHES BY THE AUTHOR.

———◆———

BOSTON.
1878.

BOSTON:

ALFRED MUDGE AND SON, PRINTERS.

PREFACE.

In the following pages I have endeavored to describe the incidents of a trip to those comparatively little-known islands, the MAGDALENS, with a view to the study of their Ornithology.

I have often thought that if collectors would give a summary of the results of their expeditions, naturalists would soon be able to estimate the geographical range of many species which at present we know very little about.

In Part Second I have given a list of the species, with notes, that were taken or observed.

Several of the specimens differed in a marked manner from their usual coloration; but while giving the points of difference in the notes regarding them, I have felt that it would be too great a risk to describe them as varieties, as of late years more than half of the "new birds" have hardly received their names before they have sunk into synonymes, thus greatly impeding science.

In conclusion, I wish to say that as every one is liable to mistakes, I hope that my readers will look leniently upon them, should they find any. I can only hope that they may prove to be "exceeding few." I would add that I have followed the classification given by Dr. Elliott Coues in his "Key."

C. B. C.

CONTENTS.

___

PART I.

	PAGE
THE MAGDALEN ISLANDS	7
BYRON ISLAND	12
BIRD ROCKS	17
A DAY'S SPORT ON GRINDSTONE ISLAND	23
HOW TO GET THERE	28

PART II.

BIRDS OF THE MAGDALEN ISLANDS	31
APPENDIX	79
INDEX	85

A NATURALIST

IN THE

MAGDALEN ISLANDS.

CHAPTER I.

THE MAGDALEN ISLANDS.

On the 17th of July, 1878, as the first streaks of light became visible in the east, I climbed to the deck of the schooner "Stewart" and saw for the first time that *terra incognita*, the Magdalen Islands, rising from the sea about ten miles to the west, and stretching northward until they faded away in a faint, misty line. What ornithologist does not feel a keen sense of pleasure upon visiting a new field of study? and as I gazed upon the gradually nearing land I experienced the feelings of a miner journeying to some supposed El Dorado, awaiting with a keen anxiety for the moment to arrive when I could begin the exploration of its treasures.

Many birds were in sight, mostly Gannets, Terns, and
Gulls, while several little Petrels were dancing over the water
in the wake of the vessel. As we approached the shore
Black Guillemots became numerous, rising from the water as
we neared them and skimming away, the white patch of their
wings showing clearly in contrast with their black bodies.
As we sailed slowly past, within a few hundred yards of Entry
Island, I was struck with the barren aspect of the shore.
Cliffs from forty to fifty feet in height, composed of red
sandstone, rose almost perpendicularly from the water's
edge, contrasting strangely with the verdure growing upon
their summits. To our left, situated in a little valley, we
could plainly see Amherst, the largest village on the islands,
its white houses looking very picturesque as they stood out
in relief against the background of green hills.

Amherst is quite a thriving little place; the inhabitants
devote themselves to fishing during the summer months, and
do a very fair business. Of late years some Americans have
built a factory here, which they devote to canning lobsters.
It is in a flourishing condition, I believe, as lobsters are very
plenty and large.

Havre aux Maison, or House Harbor, is the second in size on the islands. Here we were kindly received by Mr. J. Nelson Arsenault, who did all in his power to make us comfortable. "Vous êtes chez vous, monsieur," he repeatedly observed, after we had got quietly settled in his pretty little house.

Here and at Amherst are the only places where any one can be comfortable. On all the rest of the islands it is almost impossible to sleep for the myriads of fleas and other insects which infest them; but here, where the heat rarely exceeds seventy degrees, when winter clothes are often a necessity even in July, and fresh milk and good food to be had for the asking, this quaint little village of House Harbor offers a splendid field to the naturalist or sportsman after a winter's confinement in the city.

The Magdalens are composed of ten so-called islands, although they are all, with the exception of Entry Island, Deadman's Island, Byron Island, and the Bird Rocks, connected by a narrow sand-beach, thirty-four miles in length, and extending from East Point to Amherst.

This belt of sand is covered in many places with grass,

that affords excellent feeding for the cattle. The inhabitants take advantage of this, and during the summer months turn their animals loose to shift for themselves. We often came upon large droves of sheep, accompanied by numbers of diminutive horses, a peculiar breed, which seem to be indigenous to these islands. They are strong, hardy little fellows, generally weighing from seven to eight hundred pounds. A fair one may be purchased for twenty dollars, and the price ranges from that to seventy-five.

The inhabitants are nearly all of French descent, and still speak that language, although a Parisian would probably faint away if he listened to their conversation. I was very much amused to hear such expressions as "*Faire shake hand*" or "*La boat est venu*"; but although they have adopted many English words, and have a number of peculiar *idioms*, still, any one conversant with the French language would readily understand them.

The scenery in many parts of the island is very pretty. The cliffs of soft red sandstone have been worn away in many places by the action of the sea, forming a sort of catacomb of caves, some of them extending to the depth of seventy feet. At Wolf Island, large pillars have been left standing, seemingly guarding the entrances to the numerous caves and inlets. On a calm day, a visit to Wolf Island would well repay the tourist or naturalist, as besides the beauty of the spot, the Black Guillemot (*Uria grylle*) breeds here in large numbers.

CHAPTER II.

ONE pleasant afternoon we left House Harbor and set sail for Grand Entry Harbor, where we arrived the same evening, and spent a not very comfortable night on account of the myriads of fleas, which are much more numerous here than anywhere else on the islands.   Here, after some little trouble, we procured a boat and set sail for Byron Island.   It was a beautiful day, and our little craft sailed finely.   As we passed out of the harbor I observed several American fishing schooners, their white sails forming a striking contrast to the deep red ones of the island boats.   Numbers of Gannets and Black Guillemots dotted the water in every direction, while occasionally a seal came to the surface and gazed stupidly at us as we passed.

We made the run of twenty-four miles in about five hours, and cast anchor in a little bay, protected by cliffs that rise in some places over two hundred feet in height.   In this isolated spot a few families have made their homes. The men fish during the summer, while the women do all the work about the house. It was a rather amusing sight to see a girl of fourteen hard at work chopping wood, swinging an axe with the precision of a veteran wood-cutter.

Byron Island has an area of about four square miles, and is mostly covered by a thick growth of diminutive pines. While rambling around I observed several species of birds which I did not expect to find on such a deserted spot. Among them were *Tachycineta bicolor, Cotyle riparia, Loxia leucoptera, Loxia curvirostra var. americana, Sitta canadensis,* and a few scattered Puffins (*Fratercula arctica*). Gulls and Terns were abundant, and a solitary Blue Heron (*Ardea herodias*) stalked with solemn strides through a small marsh which had been formed in a depression of the ground by the recent rains.

On the eastern shore of Byron Island a point of sand runs out into the ocean for several hundred yards, from which grand sport may be had in September; but it cannot compare with East Point, the most northern point of the connected chain of islands.

We left too early for really good shooting. Any one wishing to visit the islands for sport should be there during the month of September, when the Ducks, Geese, Teal, etc., are very plentiful. The following is a list of some of the birds we shot during the week ending August 20, which have more interest to the sportsman than to the naturalist: —

Godwits	6
Yellowlegs	64
Teal	6
Snipe	9
Sandpipers	57
Curlew	7
Plover	74
Turnstones	21
Duck	4

Small birds may be kept, by those who have not the leisure to prepare them in the usual way, by an injection of carbolic acid, the "crude" I think the best. Take a syringe and inject a small quantity into the throat and vent, introduce a small piece of cotton into the cavities to prevent the acid from staining the feathers, and the bird is ready to be placed away in its box. Only small birds should be treated in this way. They may be skinned afterwards, if desirable, but can never be made to look well. It is necessary to be very careful in using the acid, as a drop of it on the flesh hardens it and sometimes produces an ulcer; and remember that a single drop of the acid in the eye will totally destroy the sight.

MAMMALS.

There are no large animals to be found on the islands. Foxes are common, and occasionally a Silver Fox is taken, but the latter is rare. Rabbits are numerous on some of the islands. Seals are very plenty in spring, and are hunted with eagerness by the inhabitants, as the sale of their hides and oil forms one of their most important incomes. A man told me that he had killed one hundred and eleven in one day. They kill them on the floating ice, and their hunting is oftentimes attended with serious accidents. About seventy-five years ago the Walrus was common also. I was fortunate enough to obtain a very fine skull from Byron Island, and several tusks. Their bones are still found in numbers near Grand Entry and on the beach west of Amherst.

FISH.

There are a number of brooks on the islands which are filled with trout, but none of them very large. I believe the largest that I heard of being taken weighed about a pound and a half. There are also several small lakes, which the inhabitants claim contain large fish.

INSECTS.

Nearly all of the islands are infested with fleas and other small insects, which renders it very uncomfortable to camp in the open air. Mosquitoes do not appear to be very plenty, as we were troubled but little by them, and what few there were seemed to have a great respect for tobacco smoke; so by a judicious use of that narcotic we were enabled to enjoy many pleasant moonlight evenings in comparative comfort.

BIRD ROCK.

CHAPTER III.

BIRD ROCKS.

I HAD heard such fabulous stories regarding the numbers of birds that frequent the Bird Rocks that I determined on visiting them; so one morning, having a fair wind, we started, taking with us our man of all work, "Metrick," and the owner and captain of the little craft by courtesy called a boat.

The "Rocks" are situated about twelve miles to the northeast of Byron Island, and can only be visited during very calm weather, as otherwise it would be impossible to land.

3

Even when perfectly calm, it is no easy matter to pilot a boat safely among the partly submerged rocks in front of the beach. As we approached " Little Bird," which is situated about a quarter of a mile from its larger companion, it presented the appearance of being covered with snow. "Des moineaux," remarked the captain, and so it was. The rock was covered with birds, — literally packed, — hardly a spot was left uncovered. Thousands of Gannets and Puffins were perched upon every conceivable projection.

As we passed close to them, many of the Gannets, which were constantly passing and repassing from one rock to the other, sailed within a few yards of us, their black-tipped wings and pure white plumage giving them a very pretty appearance. Every few moments some one of them would suddenly close its wings and descend with the speed of an arrow, disappearing completely beneath the water, and they generally reappeared holding some unlucky fish in their beaks which had allowed itself to approach too near the surface.

Bird Rock rises in a nearly perpendicular cliff to the height of one hundred and two feet, and has about four square acres of ground upon its summit. Here for the last few years a lighthouse has been established, and the men stationed here, when they found that we wished to land, lowered a small platform by the aid of a crane and windlass for our accommodation.

By the aid of a little careful packing, we were enabled to place everything, including ourselves, upon this substitute for an elevator. The box or tray in which we were seated was about four feet square, with six-inch sides. When we were all snugly packed away, it is unnecessary to state that we were somewhat crowded; but just then a faint voice came to us with a questioning "All right?" and I had hardly waved my hand in answer, when we began to ascend. As we moved slowly up within a few feet of the cliff, we were enabled to form some idea of the immense quantities of birds which breed here. Gannets, Puffins, Razor-billed Auks, and Guillemots, together with an occasional Kittawake Gull, were sitting upon the ledges in long rows, some upon eggs, but most of them were mixed up with young ones of all shades and sizes, from those nearly able to fly to the downy little fellows apparently just from the shell. Around, above, below, everywhere in fact, the air was full of birds. Those on the ledges did not offer to fly, but sat still and stared at us stupidly. The air was tainted with the odor arising from hundreds of fish in all stages of decomposition. Such a horrible clamor was kept up, too. One can never realize the sight until it has been seen. About half-way up, the captain and our man began to tremble and turn pale. The way they hung on to the chains was a beautiful sight to see. Every few feet the chain would run over itself as it was wound up by the windlass, and in slipping back to its place it gave us a slight jerk, that caused the captain to mutter his prayers faster than ever.

The ascent took twenty-seven minutes, and it really is not a pleasant sensation being swung about over a hundred feet in the air, supported only by a rusty chain. As we neared the top, the captain kept crying out, "For God's sake, hurry!" and when we were swung over the cliff above solid ground, he gave a deep sigh and swore that if he got down alive he would never again make the ascent.

We were kindly received by Mr. Peter Whalen, the light-keeper, and after a dinner on Guillemots' eggs we made the circuit of the island, accompanied by our genial host. Only in very calm weather can boats come ashore here, as they have to run under the cliffs and land upon a small, rocky beach, the only place where it is safe for a boat to approach the island. Mr. Whalen told me that the supply-boat, which comes here twice a year, oftentimes has to wait days before she can land the stores.

I was very much pleased to observe specimens of *Sitta canadensis* and *Cotyle riparia*, the former being very tame, and making itself quite at home about the lighthouse, and to obtain a fine immature specimen of the Arcadian Owl (*Nyctale arcadia*). Mr. Whalen showed me specimens of the Black and Yellow Warbler (*Dendroeca maculosa*), Yellow-rumpled Warbler (*Dendroeca coronata*), and Richardson's Owl (*Nyctale tengmalmi var. Richardsonii*), which he stated he obtained here in June, and that many species of small birds come to the "Rock," of which, unfortunately, he did not know the names, and had not preserved the skins.

Five species of birds breed here every year, in such numbers that the fishermen used to gather their eggs by the barrel until the government put a stop to it. These five species, the Gannet (*Sula bassana*), Common Guillemot (*Lomvia troile*), Razor-billed Auk (*Utamania torda*), Kittawake Gull (*Larus tridactylus*), and the Puffin (*Fratercula arctica*), with the exception of the latter, which generally breeds in a hole, rear their young side by side, and appear to live as amicably as possible.

CHAPTER IV.

A DAY'S SPORT ON GRINDSTONE ISLAND.

LET me try to sketch a day's sport to be had on these charming little islands. To do this, I do not know that I can do better than to give an extract from my journal.

TUESDAY, Aug. 20, 1878. — As I arose and opened my window the soft notes of the Blackbreast Plover were borne faintly to me by the morning breeze. The air was crisp and cold; and although the sun was shining brightly, a strong wind still continued, making it impracticable to try and reach the shooting-ground by boat; so after a hearty breakfast we harnessed up our little horse and started around by road, a distance of about four miles. As we passed in and out of the clumps of woods, numbers of birds were constantly to be seen, filling the air with their cheerful little songs. Robins, Snowbirds, and Blackpoll Warblers appeared to be the most common species. Numbers of Sparrows were incessantly

darting in and out of the undergrowth, while an occasional
Crow sailed slowly across the road, until perceiving us he
would hurry away, uttering a parting "caw" of joy at his
supposed lucky escape. Curlew were very abundant, but
shy, as they always are. At times, large flocks would rise
from the fields and fly silently for a few hundred yards
before again alighting in some field more secure from obser-
vation; and sometimes, although none could be seen, their
soft, trilling whistle would be heard, as though some watch
ful sentinel was giving notice to his companions of our
whereabouts.

As we crossed the beach to our "stands" large flocks of
Plover and Yellowlegs started up in front of us, but they did
not fly far, the wind being too strong for them to remain
long exposed to it.

Hardly had I got fairly settled in my box before a flock of
about a dozen Blackbreast Plover settled among my decoys,
from which I bagged four with the first barrel, and another
with the second. I had just picked them up, and was
returning to the "stand," when a Hudsonian Godwit lit
among the decoys without paying the slightest attention to

me. He apparently came to the conclusion that they were not of his own kind, for after looking around for a moment he started away, passing almost directly over my box, into which he nearly fell when I shot him. Yellowlegs (*Totanus melanoleucus*) were very abundant, and I was amused to see my friend kill five young birds of that species, one after another, which allowed him to walk within shot of them.

The marsh in which I had placed my " box " extended in an unbroken line for nearly twelve miles, and is sheltered on either side by sand-hills that rise from twenty to fifty feet in height. Here the birds love to feed, especially on a stormy day, as they are protected from the wind by the sand-hills, and find plenty of food among the rich grass, and in the large, shallow pools of brackish water.

Sandpipers in large flocks were constantly passing and repassing; in fact, I found the shooting here to have all the pleasure of deep-sea fishing, in the uncertainty of what would next be taken.

I had just shot a fine specimen of the great Blue Heron (*Ardea herodias*), when a flock of Teal passed on one side of me, while a flock of about twenty Plover were passing on

4

the other. In trying to do too much, I did nothing; for I
only obtained one Teal. Later in the fall, this is a grand
place for Geese, which come in here to drink and wash themselves in the numerous pools of partly fresh water.

On this day we made quite a varied bag, which I give
in detail: —

Hudsonian Curlew	1
Hudsonian Godwits	2
Blackbreast Plover	20
Golden Plover	2
Semipalmated Plover	9
Turnstones	5
Greater Yellowlegs	12
Yellowlegs	9
Least Sandpipers	7
White-rumpled Sandpipers	13
Pectoral Sandpiper	1
American Dunlin	1
Semipalmated Sandpipers	5
Wilson's Snipe	2
Red-breasted Snipe	4
Teal	1
Blue Heron	1
	95

Ravens were very common. I must have seen over a dozen of them. Gulls were also numerous, mostly in large flocks, as if preparing to go south.

On the way back I observed a Short-eared Owl (*Brachyotus palustris*) flying low over a field, but was unable to obtain a shot at it.

We were very hungry, and were glad enough to find that our hostess had been on the watch for us, and had dinner all ready upon our arrival. Any one who knows how hungry a man can get after a hard day's shooting will appreciate, as we did, the dinner which we sat down to, beginning with Plover and Curlew, cooked to a turn, boiled potatoes, corn bread, white bread and jelly, and ending with an old-fashioned plum pudding, with hot coffee and cold spring water to wash them down.

After dinner it was our custom, in fine weather, to roll into the hammocks in front of the house and enjoy our cigars in the moonlight, while the only sounds that broke the peaceful silence of the scene were the occasional "goodnight" notes of the Golden-crowned Thrush and the indistinct sound of the waves breaking upon the beach.

CHAPTER V.

HOW TO GET THERE.

THERE is an old saying that " Knowledge can do no harm," and therefore I do not think that a few remarks on the ways and means of getting to the islands will be out of place here.

From St. John, N. B., to Pictou, N. S., is a ride of about sixteen hours, and tiresome enough it is, too; leaving St. John at half past eight in the morning, and arriving at Pictou at about ten o'clock the same evening. The best hotel in Pictou is the St. Lawrence House.

From Pictou the steamer "Albert" runs twice a month to the Magdalen Islands, in fine weather touching at Georgetown and Souris, P. E. I. Another way of getting there, for those who have plenty of time at their disposal, is to go to Souris, P. E. I., and charter a schooner, or stop there and wait for the "Albert." Souris offers some fair shooting in the season, and also some very good trout fishing.

Good accommodations can be obtained at the Bay View House, owned by James McDonald, Esq. No delicacies can be obtained there. Such things as good coffee, etc., should be purchased in St. John.

On the Magdalens, at Amherst Harbor, will be found plenty of quaint little places to stop at. Those kept by Mrs. Burns and Mrs. Shea are to be recommended.

At House Harbor, Mr. J. Nelson Arsenault will do all in his power to make any one comfortable. Mr. Boudrou will also take a few boarders.

On all the other islands the traveller must depend upon the hospitality of the inhabitants.

PART II.

CATALOGUE

OF

BIRDS TAKEN OR OBSERVED

IN THE

MAGDALEN ISLANDS,

WITH

NOTES REGARDING THOSE FOUND BREEDING,

ETC, ETC.

CATALOGUE.

---

1. *TURDUS MIGRATORIUS, L.*
(Robin.)

This is one of the most abundant birds on the islands. Many of the specimens taken varied greatly in coloration and size.

It is just possible that some few birds may remain all winter, as I was told by the inhabitants that they often observed them as late as December.

Breeds.

2. *TURDUS PALLASI, Cab.*
(Hermit Thrush.)

The present species was not uncommon. I procured specimens of both young and old birds.

Breeds.

5

3. *TURDUS SWAINSONI,* Cab.

(Olive-backed Thrush.)

I do not think it is at all uncommon, although but one specimen was taken, an adult male.

Probably breeds.

4. *MIMUS CAROLINENSIS,* (L.) Gr.

(Catbird)

I include this species with hesitation, as we did not take a single specimen; but upon two occasions I heard what I believed to be its peculiar cry.

5. *SAXICOLA ŒNANTHE,* (L.) Bechstein.

(Stone Chat.)

On August 22, while walking along a narrow roadway, a bird suddenly started up in front of me and lit on a fence a few yards away. It was very tame, and allowed me to approach within a few feet of it, while it sat cocking its head knowingly from side to side. Unfortunately, I had no gun; but there was no mistaking a bird so well known to me as *Saxicola œnanthe*, having taken a series of over fifty specimens in Egypt, where it is very abundant.

It can only be considered as a straggler, although " Dr. H. R. Storer found them breeding in Labrador, in the summer of 1848, and procured specimens of the young birds." (Vol. I, p. 60, Bd. Bwr. Ridgw. Bds. N. America.)

6. *REGULUS SATRAPA, Licht.*
(GOLDEN-CRESTED KINGLET.)

Quite common. Several were taken at House Harbor, including one immature specimen.

Breeds.

7. *PARUS ATRICAPILLUS, L.*
(BLACK-CAPPED CHICKADEE.)

Abundant in the thickly wooded parts of the islands. Probably breeds.

8. *PARUS HUDSONICUS,* Forster.
(Hudsonian Titmouse.)

Not uncommon, but does not appear to be as plenty as the last species. We found it quite abundant at House Harbor.

9. *SITTA CANADENSIS, L.*
(Red-bellied Nuthatch.)

Appears to be common on some of the islands. At Bird Rock it came into the lighthouse without the least signs of fear, and began hopping around, looking for its food, without paying the slightest attention to us. I also observed several on Byron Island.

Breeds.

10. *ANORTHURA TROGLODYTES,* (*L.*) *Cs.*
var. hyemalis, (*Wils.*) *Cs.*
(Winter Wren.)

Rather common. Killed three, and observed several others.

11. *EREMOPHILA ALPESTRIS,* (*Forst.*) *Boie.*
(SHORE LARK.)

I procured a single specimen from Mr. William Perham, which he had killed on Byron Island. It was a fully fledged young bird, still retaining the spotted back. No others were observed during our stay on the islands.

Breeds.

12. *MINOTILTA VARIA,* (*L.*) *V.*
(BLACK-AND-WHITE CREEPER.)

My first Creeper was taken on July 27, and I observed them afterwards occasionally. It cannot be considered an abundant species.

Probably breeds.

13. *DENDRŒCA ÆSTIVA,* (*Gm.*) *Bd.*
(SUMMER WARBLER.)

Common during the month of July; hardly a day passed without our seeing one or more of them; but after August 15 they became quite scarce.

Breeds.

14. *DENDRŒCA CORONATA*, (L.) *Gr.*

(YELLOW-RUMPLED WARBLER.)

Abundant on all the islands, especially so at House Harbor.

Breeds.

15. *DENDRŒCA STRIATA*, (*Forst.*) *Bd.*

(BLACK-POLL WARBLER.)

Very common until September. We procured a number of specimens of young birds.

Breeds abundantly.

16. *DENDRŒCA MACULOSA*, (*Gm.*) *Bd.*

(BLACK-AND-YELLOW WARBLER.)

While at Bird Rock I was shown a mummified specimen of the present species. It had been killed, and kept as being a "pretty bird"; and having no means of preserving it, they had simply injected it with salt and hung it up to dry. It was taken in June.

17. *SEIURUS AUROCAPILLUS*, (*L.*) *Sw.*
(Golden-crowned Thrush.)

Unfortunately, I was unable to obtain a single specimen, but they appear to be quite common. Often, on pleasant evenings, I have sat and listened to their almost unmistakable notes, and counted as many as four of them singing in one small patch of woods.

Probably breeds.

18. *SEIURUS NOVEBORACENSIS*, (*Gm.*) *Nutt.*
(Water Thrush)

Procured the first specimen July 19, and another a few days later. Both were males.

Probably breeds.

19. *SETOPHAGA RUTICILLA*, (*L.*) *Sw.*
(Redstart.)

Very plentiful all over the islands wherever woods are to be found. They begin to leave for the South about August 18.

Breeds.

20. *TACHYCINETA BICOLOR*, (*V.*) *Cs.*
(WHITE-BELLIED SWALLOW.)

Observed it occasionally during the latter part of the month of July.   The last specimen procured was shot August 6.   Does not appear to be very abundant.

21. *COTYLE RIPARIA*, (*L.*) *Boie.*
(BANK SWALLOW.)

Very abundant everywhere from Bird Rock to Amherst. Breeds.

22. *AMPELIS CEDRORUM*, (*V.*) *Bd.*
(CEDAR BIRD.)

Not at all common.   Observed a small flock on July 26, and another August 6.   On August 21 I killed a young bird hardly able to fly.

Breeds.

23. *PINICOLA ENUCLEATOR,* (L.) V.

(PINE GROSBEAK.)

A rather scarce bird on the islands. Mr. Perham shot one in July, and I observed two a few days later and took one August 9.

Probably breeds.

24. *CARPODACUS PURPUREUS,* (Gm.) Gr.

(PURPLE FINCH.)

Did not find it common. One specimen was taken August 3, and another on the 5th.

25. *LOXIA LEUCOPTERA,* (Wils)

(WHITE-WINGED CROSSBILL.)

A common species. I found it abundant in the thick pine woods of the southern islands. Several full-grown young and adults were taken in July. It probably breeds here early in the spring. It has a pleasing song.

26   *LOXIA CURVIROSTRA var. AMERICANA, (Wils.) Cs.*

(COMMON CROSSBILL.)

Abundant, and very tame in the woods near House Harbor. I also procured a specimen on Byron Island. Breeds.

27.   *ÆGEOTHUS LINARIA, (L) Cab.*

(RED-POLL LINNET.)

I include the present species with great hesitation. On August 3 a bird was brought to me, in a terribly mangled condition, which I believed to be *Ægeothus linaria.* No others observed or taken.

28.   *CHRYSOMITRIS PINUS, (Wils) Bp.*

(PINE LINNET.)

A single specimen, taken July 30, which proved to be an adult male.

29. *PASSERCULUS SAVANNA, (Wils.) Bp.*

(SAVANNA SPARROW.)

An abundant species. It was very common at House Harbor, where I procured a fine series of specimens.
Breeds.

30. *POOECETES GRAMINEUS, (Gm.) Bd.*

(BAY-WINGED BUNTING.)

A rather common species, but not nearly as numerous as the last.
Probably breeds.

31. *MELOSPIZA PALUSTRIS, (Wils.) Bd.*

(SWAMP SPARROW)

I believe it to be common, although only four specimens were obtained; but this would be accounted for by their retiring habits. I have often nearly stepped upon them before they would take to flight.
Breeds.

32. *MELOSPIZA MELODIA, (Wils) Bd.*

(Song Sparrow)

It cannot be considered as common; only two specimens were taken, but I have no doubt that it is a regular summer visitor, and breeds here.

33. *JUNCO HYEMALIS, (L.) Scl.*

(Snowbird.)

An abundant species. They were very tame, and came around the house fearlessly. On July 18 I discovered a nest containing young birds nearly able to fly. The old birds showed their anxiety as I approached by flitting from one branch to another, sometimes almost within reach of my hand, continually uttering a short, clear "chip" of alarm.

In a series of specimens measured, I found quite an appreciable difference in size; some adult birds were darker than others.

34. *SPIZELLA SOCIALIS,* (*Wils.*) *Bp.*

(Chipping Sparrow.)

One specimen taken. It may have been a straggler, but I am inclined to think that a few pairs breed on the islands every year.

35. *ZONOTRICHIA ALBICOLLIS,* (*Gm.*) *Bp.*

(White-throated Sparrow.)

A common summer visitor. It was abundant on Grindstone and Amherst Islands.

Oftentimes, after a long tramp, I have rested myself under some tree, and there, in the concealment of its drooping branches, have listened with pleasure to two or three of these birds piping their pretty little whistles, one after the other, sometimes each in a different key. Whistle three notes of a chord, beginning with the lowest, and repeating the last three or four times; some idea of their notes may thus be obtained; but only those who have had the pleasure of hearing this little chorister in its native freedom can appreciate the sad and plaintive softness of its simple little song.

It breeds commonly.

36. *ZONOTRICHIA LEUCOPHRYS*, (*Forst.*) *Sw.*
(WHITE-CROWNED SPARROW.)

The present species cannot be uncommon during its migrations to and from Labrabor.

37. *PASSERELLA ILIACA*, (*Merrem*) *Sw.*
(FOX SPARROW)

A common species. Unfortunately, I arrived too late to find it breeding, but I have no doubt that it does so. The first specimen was taken July 19, and I killed ten within the next two weeks. All of them, with one exception, had the wing-marking more or less distinct, but several of them varied perceptibly in coloration.

The other specimen, which was taken August 12, approached very nearly the description of the so-called *Passer-*

ella obscura, Verrill, but I believe it to be simply a stage of plumage of *Passerella iliaca.*

Breeds.

38. *SCOLECOPHAGUS FERRUGINEUS*, (*Gm.*) *Sw.*

(Rusty Grackle)

Abundant. On August 15 I observed large flocks of them, apparently preparing to leave for the South.

Breeds.

39. *CORVUS CORAX, L.*

(Raven.)

An abundant species. They were shy, but hardly more so than the Crow is in Massachusetts. The first specimen was taken July 27, and another August 3. Hardly a day passed that I did not see two or three of them, either perched upon the cliffs or flying high up in the air, uttering their dismal croaks.

A pair bred this spring in an almost inaccessible spot, high upon the cliffs of Cap au Meule, and another near Hospital Island.

40. *CORVUS AMERICANUS, Aud.*

(Common Crow.)

Very abundant, and quite tame in comparison with their usual shyness. One afternoon, while making my way through some thick brush, I heard a number of them within a short distance. Concealing myself under a tree, I commenced to call, and in a few minutes several came within shot, and circled above me until I had killed four.

The hatred of a Crow for all species of Owls is a well-known fact, and is sometimes taken advantage of to entrap them. Mr. Perham related to me a curious manner of waging war on the "black robbers" which I have never seen attempted anywhere else. He places a live Owl in some isolated tree near where the Crows feed, and concealing himself within easy shot, imitates the hoarse cry of a Crow, upon the discovery of his enemy, to perfection. Generally, the cry is taken up and the Crows come from all directions. "The more you kill the madder they get," said he, "but it is necessary to keep concealed." From forty to fifty Crows have been taken in this manner in one afternoon.

Breeds commonly.

41. *EMPIDONAX TRAILLII, (Aud) Bd.*
(Traill's Flycatcher.)

On August 15, while walking along the edge of a small swamp, a bird suddenly flew past me and lit on a branch a few yards away. So eager was I to obtain it, as I had recognized it to be a Flycatcher, that I fired instantly, and blew the poor little fellow almost to pieces. Upon examination, I decided it to be the present species; still, *Empidonax traillii* and *Empidonax minimus* so much resemble each other that a badly mangled specimen of either is difficult to identify.

42. *EMPIDONAX MINIMUS, Bd.*
(Least Flycatcher.)

A common species, and not at all shy. While at House Harbor a girl brought me one which she had caught alive in the barn. It did not appear to be at all hurt, so after examining it I opened my hand and gave it freedom. I was leaning against a fence at the time, and upon being released it simply hopped from my hand to the top rail of the fence and sat there quietly, allowing me to stroke it gently with my finger, all the while looking at me knowingly. Perhaps some movement on my part startled it, for it suddenly flew away and disappeared in the brush.

Breeds.

43.　*EMPIDONAX FLAVIVENTRIS, Bd.*
(Yellow-bellied Flycatcher)

Procured two specimens, but it does not appear to be at all common.

Probably breeds.

44.　*CERYLE ALCYON, (L.) Boie.*
(Belted Kingfisher.)

The Belted Kingfisher appears to be as abundant on the islands as it is almost everywhere else. Hardly a pool of fresh water was seen that did not have one or more of this species either hovering above it or perched upon some projecting branch that commanded a view of its surface.

Breeds.

45. *COCCYZUS ERYTHROPHTHALMUS*, (*Wils.*) *Bd.*

(BLACK-BILLED CUCKOO.)

A single specimen taken at House Harbor, July 19, which proved to be an adult male. I saw no other while on the islands.

Probably breeds.

46. *PICUS VILLOSUS, L.*

(HAIRY WOODPECKER.)

The first specimen was taken August 5. I only observed it on four occasions; obtained one bird of the year.

Breeds.

47. *PICUS PUBESCENS, L.*

(DOWNY WOODPECKER.)

Not at all common. I observed one specimen August 3, and shot another August 20, at House Harbor.

Breeds.

48. *COLAPTES AURITUS*, (L.) *Sw.*

(Golden-winged Woodpecker.)

Abundant all over the islands. It is called " Pivert " by the islanders.

Breeds.

49. *OTUS VULGARIS var. WILSONIANUS*, (*Less*) *All.*

(Long-eared Owl.)

A specimen of this bird was killed at House Harbor July 20, and another July 27. I observed it occasionally afterwards.

Breeds.

50. *BRACHYOTUS PALUSTRIS*, *Auct.*

(Short-eared Owl.)

Upon two occasions I saw what I believed to be this bird on Hospital Island, and Mr. Perham showed me a specimen which he had killed at House Harbor. He had also observed others.

Breeds.

51. *SYRNUIM NEBULOSUM*, (*Forst.*) *Gr.*
(Barred Owl)

Either this species or *Syrnium lapponicum var. cinereum* is common late in the fall, perhaps both. The inhabitants state that immense grayish-looking Owls were abundant at times. I was unable to find either species.

52. *NYCTEA NIVEA*, (*Daud.*) *Gr.*
(Snowy Owl.)

A very common spring and autumn visitor. I saw one on Byron Island, August 1, and two or three afterwards. The inhabitants state that they are found on the beaches in large numbers late in the fall.

53. *SURNIA ULULA var. HUDSONICA*, (*Gm.*) *Ridge.*
(Hawk Owl.)

A common species. One specimen observed July 28, another taken August 5, and a fine adult male August 23. All were taken during the daytime.

Breeds.

54. *NYCTALE TENGMALMI var. RICHARDSONII,* (*Bp*) *Ridge.*
(Richardson's Owl.)

Mr. Whalen showed me a specimen which he had killed in June, and stated that he had seen three more this year. I found it quite common at House Harbor, where I shot one specimen and heard several others.

Mr. William Perham showed me an egg of this species which he had taken on June 14 at House Harbor. The nest was a hole in a dead birch, and contained an addled egg, and four young birds nearly able to fly. The young were much darker than the old birds, both of which were procured. The egg, which is now in my collection, measures 1.6 inches in length by 1.25 in breadth. It is of a creamy white color.

55. *NYCTALE ACADICA,* (*Gm.*) *Bp.*
(Arcadian Owl.)

On July 30 I shot a single specimen of this little Owl on Bird Rock. It was a young bird, and had the brown markings and white face which once was supposed to represent

Nyctale albifrons. The taking of this young bird renders it probable that *Nyctale acadica* breeds on the islands; but it is not abundant, as no others were taken.

A few days later, on Byron Island, while taking an evening stroll, a small Owl circled silently around me for several minutes, which, from the momentary glimpses obtained of it in the uncertain light, I believed to be the present species.

56. *CIRCUS CYANEUS var. HUDSONICUS,* (L.) *Cs.*
(MARSH HAWK.)

A rather common species. I saw it several times at House Harbor, and once on Byron Island. Mr. Perham killed a specimen on Bird Rock.

Probably breeds.

57. *ACCIPITER FUSCUS,* (Gm.) *Bp.*
(SHARP-SHINNED HAWK.)

Observed what I believed to be this Hawk at Hospital Island, but was unable to approach within shot of it.

Obs. *FALCO COMMUNIS, Variorum.*

(Peregrine Falcon.)

A fisherman told me that last year he shot a Hawk which had "killed a small Duck," and which "had a black back and spotted breast." It was probably the present species, and there is no reason why it should not visit here. I have taken it on Prince Edward's Island.

58. *FALCO COLUMBARIUS, L.*

(Pigeon Hawk.)

One was seen at House Harbor and another at Amherst.

On July 10 Mr. Perham found a nest of this species on Byron Island. It was built in a fir-tree, and contained two fresh eggs. He did not take the bird. The inhabitants stated that it breeds here every year, and that they sometimes found nests containing eight or nine eggs. The latter statement I do not think correct, as the number rarely exceeds five, according to all accounts.

59. *AQUILA CHRYSAETUS, L.*
(Golden Eagle.)

On August 16 some fishermen killed a large Eagle at Amherst. From the description of the bird, I believe it to have been *Aquila chrysaetus*, although it might have been a young *Heliaetus leucocephalus*.

60. *LAGOPUS ALBUS, (Gm.) Aud.*
(Willow Ptarmigan.)

In 1876 a single specimen was taken at Amherst. The skin was preserved. It is, I believe, the only example of this species ever taken on the islands, and was probably a straggler from Newfoundland. I was unable to obtain the exact date of its capture.

61. *SQUATAROLA HELVETICA, (L.) Cur.*
(Black-bellied Plover.)

The first specimens were seen August 10, but were very shy. I procured two on August 11, and after that date they were very common and quite tame.

8

62. *CHARADRIUS FULVUS var. VIRGINICUS,* Borck.

(Golden Plover)

Arrives about August 11, and is abundant after August 15. The inhabitants state that it is common here until October.

The American Golden Plover is known to be very closely allied to *Charadrius fulvus,* of Asia, so closely in fact that in some cases it is almost impossible to distinguish them. As regards *Virginicus,* it has the gray axillaries, under wing coverts, and slightly feathered tibia, of *Charadrius fulvus,* and the only difference that is perceptible is a slight variation in size. If this difference in size should be constant, then *Charadrius virginicus* would constitute a good variety if not a species.

In a large series of Golden Plovers from Asia and America, as well as from other parts of the world, which it has been my good fortune to examine, I do not find any greater difference in size than often exists in individuals of the same species of some of our common Sandpipers, which no one for a moment would think of separating as varieties or species on that account.

Mr. Dresser states ("Birds of Europe," Part IX) that he has a specimen of *Charadrius fulvus* in his collection "which was taken at sea in lat. 69° 30′, N. long. 173° 20′ E., many miles northwest of Point Barrow," and "much nearer to the American than to the Asiatic coast." He also states that " Mr. Pickering, when at sea on the 13th of November, between the Sandwich Islands and California, procured specimens which were evidently migrating."

The following list gives the comparative measurements of Golden Plovers taken in different parts of the world. Those which are not my own are taken from Mr. H. E. Dresser's " Birds of Europe ":—

CHARADRIUS FULVUS.

	LENGTH.	WING.	TAIL.	TARSUS.	MI . TOE
N. E. AFRICA :					
Djedda,	8.3	6.2	2.4	1.6	1.05
Alexandria,	8.5	6.25	2.25	1.55	1.0
SIBERIA :					
Lake Baikal,	8.3	6.25	2.5	1.55	.95
INDIA :					
Indian Peninsula,	8.5	6.25	2.25	1 55	1.0
CEYLON :					
Oripo,	9.0	6.1	2.4	1.55	1.5

	LENGTH.	WING.	TAIL.	TARSUS.	MID. TOE.
MALACCA,	9.0	6.25	2.35	1.55	1.0
JAVA,	9.0	6.4	2.4	1.55	1.0
BANKA,	8.0	6.1	2.3	1.5	1.0
BORNEO,	8.9	6.05	2.4	1.6	1.05
BALCHIAN,	9.5	6.2	2.5	1.6	.95
AUSTRALIA:					
Queensland,	9.0	6.55	2.5	1.6	1.0
CHINA:					
Foochow,	9.0	6.4	2.4	1.6	1.0
"	8.5	6.25	2.25	1.55	1.0
"	7.5	5.9	1.9	1.5	0.9

CHARADRIUS VIRGINICUS.

	LENGTH.	WING.	TAIL.	TARSUS.	MID. TOE.
HUDSON'S BAY:					
Moose Factory,	8.8	6.6	2.5	1.7	0.9
TEXAS,	9.0	6.6	2.7	1.7	1.0
GUATEMALA,	9.0	7.0	2.9	1.7	0.9
PERU:					
Yambo Valley,	9.0	6.7	2.6	1.6	0.9
FLORIDA:					
Pilot Town,	9.0	6.8	2.6	1.7	0.9
MASSACHUSETTS:					
Chatham	9.1	6.8	2.8	1.7	.95
"	8.9	6.6	2.6	1.65	.9
PRINCE EDWARD'S ISLAND:					
Malpeque,	8.9	6.7	2.7	1.7	.9
MAGDALEN ISLANDS,	9.0	6.7	2.8	1.7	.9

Although there appears to be a constant difference in size,* yet it will be seen by the preceding list that specimens of the same species differ as much in size as the so-called *var. Virginicus* differs from *Charadrius fulvus.*

63. *ÆGIALITIS SEMIPALMATUS,* (*Bp.*) *Cab.*
(Ringneck.)

An abundant species during the migrations. Some probably breed, as I found it common in July.

64. *ÆGIALITIS MELODUS var. CIRCUMCINTA,* *Ridg.*
(Piping Plover.)

An abundant species. I procured immature birds in July, and Mr. Perham found it breeding in June. This, I believe, is the most northern record of its capture, but future research, I think, will reveal that it has a much more northern range than naturalists generally suppose. Dur-

---

* Not greater than would be caused by climatic influences and geographical distribution.

ing the month of August I procured specimens which I think may have been migrating. If this should prove true, it is possible that its range may extend to *Anticosti*, *or even to Labrador.*

65. *STREPSILAS INTERPRES*, (*L.*) *Ill.*
(Turnstone.)

Very abundant after July 30. Procured one specimen July 26.

66. *GALLINAGO WILSONI*, (*Temm.*) *Bp.*
(Wilson's Snipe.)

Abundant; especially so on Red Cape Marshes. Breeds.

67. *MACRORHAMPHUS GRISEUS*, (*Gm.*) *Leach.*
(Red-breasted Snipe.)

Common during the migrations. Procured the first specimen August 11.

68. *EREUNETES PUSILLUS*, (*L.*) *Cass.*
(SEMIPALMATED SANDPIPER.)

Abundant during the migrations.   First taken July 21.

69. *TRINGA MINUTILLA*, *V.*
(LEAST SANDPIPER.)

Abundant after July 21.   I saw some flocks that must have contained several hundred individuals.

It is pleasant to note the utter fearlessness of this graceful little species, when unacquainted with their greatest enemy, man.   Oftentimes while reclining upon the beach I have watched these little creatures running from side to side, searching diligently for their food, at times approaching within a few feet of me.   Upon my making a movement, they showed no signs of fear otherwise than by taking two or three little running side steps to slightly increase the distance between us, and then continued their earnest search, peeping softly to themselves, happy in the present, seemingly without a presentiment of the dangers of their coming journey south.

70. *TRINGA MACULATA, V.*
(Pectoral Sandpiper.)

I shot a single specimen August 20, and another on the 21st; but it is probably common during the migrations.

71. *TRINGA BONAPARTEI, Schl.*
(White-rumpled Sandpiper.)

Abundant everywhere. At times I have seen the beaches covered with them. They arrive about August 10.

72. *TRINGA ALPINA var. AMERICANA, Cass.*
(American Dunlin.)

Common during the migrations. Arrives about the middle of August.

73. *TRINGA CANUTUS, L.*
(Red-breasted Sandpiper.)

Arrives about August 10. Procured only two, but it is probably common during the migrations.

74. *CALIDRIS ARENARIA, (L.) Ill.*
(SANDERLING.)

First taken August 11, afterwards became rather common.

75. *LIMOSA HUDSONICA, (Lath.) Sw.*
(HUDSONIAN GODWIT.)

First obtained August 8. An abundant species during the migrations, and exceedingly tame. Upon several occasions I have "walked them up" and shot them.

Audubon states that he was informed that *Limosa hudsonica* was to be found breeding in the Magdalen Islands, and that his friend Thomas McCulloch confirmed that statement, and also informed him that these birds breed at times on Prince Edward's Island. At the present time none remain to breed so far south.

76. *TOTANUS MELANOLEUCUS, Gm.*
(GREATER TELL-TALE.)

Arrives about July 15, and afterwards becomes very abundant.

9

77. *TOTANUS FLAVIPES, Gm.*

(YELLOW-SHANKS.)

Very common during the migrations.   Arrives about
July 16, and becomes abundant about July 26.

78. *TRINGOIDES MACULARIUS, (L.) Gr.*

(SPOTTED SANDPIPER.)

Common during the summer months.   It was quite tame.
Mr. Perham found it breeding on Byron Island.
Breeds.

79. *NUMENIUS LONGIROSTRIS, Wils.*

(LONG-BILLED CURLEW.)

On August 12 I observed two birds which I believed to
be of this species.

One day I showed a specimen of the Hudsonian Curlew
to an old seal hunter and asked him if he had ever killed
any larger.   He answered that he had occasionally killed
them larger, "*avec le bec beaucoup plus longue.*"

80. *NUMENIUS HUDSONICUS, Lath.*
(HUDSONIAN CURLEW.)

Abundant during the migrations. The first arrive about July 24, and soon after become very numerous. The fields along the road to Hospital Island are filled with berries, and is a great resort for them. I have started as many as eleven flocks in one afternoon, some of them numbering hundreds of birds.

81. *NUMENIUS BOREALIS, (Forst.) Lath.*
(ESQUIMAUX CURLEW.)

Observed a pair of these birds feeding in a field, but was unable to obtain a shot. I pointed them out to a fisherman, who seemed to recognize them, and stated that they were rare visitors. Although no others were seen, I believe that they regularly visit the islands during the migrations, and are abundant in September. We found them common near Malpeque, P. E. I., in 1875.

82. *ARDEA HERODIAS, L.*
(GREAT BLUE HERON.)

A common summer visitor. Breeds.

83. *BOTAURUS MINOR*, Gm.

(Bittern.)

Abundant. While Snipe shooting near Red Cape, I have seen as many as six of them in one afternoon. One can hardly realize the strength which lies in the slender neck of an old Bittern. The long, sharp beak, almost as keen as a knife, is capable of inflicting a severe and painful wound, as I know to my cost, having once incautiously allowed a wounded bird to strike my hand.

I was told a little incident which may illustrate the deadly use this weapon of defence is sometimes put to. One afternoon a gentleman saw a Hawk dart suddenly down into a marsh near his house. Seizing his gun, he ran towards the spot in hopes of obtaining a shot. He continued on until he suddenly came upon the dead bodies of the Hawk and an old Bittern. The beak of the latter was driven quite through the body of the Hawk, whose talons had crushed out its life, but not before it had been able to avenge itself.

Breeds.

84. *BRANTA BERNICLA, L.*
(BRANT GOOSE.)

Abundant in the fall and spring during the migrations.

85. *BRANT CANADENSIS, L.*
(CANADA GOOSE.)

The first was seen August 22. They are very abundant in the spring and fall, when the inhabitants, who do not deign to shoot smaller birds, load their muskets with buckshot and repair to the ponds, where they kill great numbers of them. I was told that twenty years ago they bred abundantly at East Point, but of late years none have been observed.

86. *ANAS OBSCURA, Gm.*
(DUSKY DUCK.)

Common all over the islands wherever fresh-water ponds are to be found.

Breeds.

87. QUERQUEDULA CAROLINENSIS, Gm.
(GREEN-WINGED TEAL.)

Very abundant. While shooting over the Three Pond
Marshes I witnessed the curious actions of a wounded bird
of this species, called " towering," by sportsmen, I believe.

While passing through a small bit of marsh bordering one
of the ponds, a pair of Teal started up in front of me. The
first dropped dead, and the other, upon receiving the con-
tents of the second barrel, rose straight up into the air,
higher and higher, turning round and round like a top,
the wings keeping up a spasmodic kind of motion, until at
an immense height it suddenly ceased all movement, and
turning over and over in its descent, struck the water a dead
weight, and when I reached the spot was stone dead.

Possibly breeds.

88. QUERQUEDULA DISCORS, (L.) Steph.
(BLUE-WINGED TEAL.)

Common, but not so plentiful as the last species.
Breeds.

89. *FULEGULA MARILA,* (*L.*) *Steph.*
(GREATER BLACKHEAD.)

On July 24 I killed a fine specimen, an adult male. He was accompanied by another, possibly a female, which I also brought down, but which was only wounded and escaped.

90. *HERALDA GLACIALIS,* (*L.*) *Leach.*
(LONG-TAILED DUCK)

. Procured a single specimen August 14. Very abundant during the migrations.

91. *HISTRIONICUS TORQUATUS,* (*L.*) *Bp.*
(HARLEQUIN DUCK.)

I was told by an old gunner that a Duck answering the description of *Histrionicus torquatus* was occasionally killed here late in the fall.

92. *SOMATERIA MOLLISSIMA,* (*L.*) *Leach.*
(EIDER DUCK.)

A skin was shown me at Bird Rock. Abundant during the migrations.

93. *MERGUS SERRATOR, L.*
(Red-breasted Merganser.)

A rather common summer resident. While walking along the shore of a small inlet we suddenly came upon a female with a brood of young birds. Upon perceiving us she seemed greatly troubled, and swam away, closely followed by her offspring, with the exception of one weakly little fellow which was apparently unable to keep up. Observing this, the old bird swam back and seemed to give some directions to the little one, for he scrambled upon her back and nestled snugly down behind her neck. As soon as he had fairly settled himself comfortably, away she went, this time closely followed by her brood. We watched her until she disappeared around a point, unharmed, and bearing the little one still upon her back, to a place of safety.

Breeds.

94. *SULA BASSANA, L.*
(Gannet.)

Thousands of these birds breed upon Bird Rock each year. They are in such numbers that oftentimes it gives the rock the appearance of being covered with snow. While on Bird Rock I had the pleasure of seeing them diving at a school of fish. It appeared to me as if some one was pouring beans from a pail, so thickly did they dive. Thousands of them seemed to go down in one continuous stream. Each bird must consume a wonderful amount of fish in a year. I have taken as many as six small mackerel out of the stomach of one Gannet. While flying, they sometimes emit a cry resembling a harsh laugh.

95. *GRACULUS CARBO, (L.) Gray.*

(COMMON CORMORANT.)

Common, and is said to breed on some of the islands.

96. *LARUS MARINUS, L.*

(GREAT BLACK-BACKED GULL.)

First observed July 21, and became common in August, collecting together in large flocks.

10

97. *LARUS ARGENTATUS, Brunn.*

(Herring Gull.)

A fine specimen was taken July 27. Later, it became common, but was not as abundant as the last species. Both this species and *Larus marinus* are eaten by the inhabitants and are considered a great delicacy. In fact, a young bird made into a stew is really quite a palatable dish, as I can testify by experience.

98. *LARUS TRIDACTYLUS, L.*

(Kittiwake Gull.)

One of the five species found breeding upon Bird Rock. The eggs are generally very difficult to procure, as they are usually deposited upon almost inaccessible ledges. They rear their young oftentimes side by side with the Gannets, Auks, and Guillemots, apparently living as amicably as possible.

99. *LARUS PHILADELPHIA, (Ord.) Coues.*
(BONAPARTE'S GULL.)

First observed August 9, and afterwards became very abundant. About August 16 they began to assemble in large flocks. This bird is sometimes eaten by the inhabitants, although the flesh is exceedingly fishy.

100. *STERNA HIRUNDO, L.*
(COMMON TERN.)

A very abundant species everywhere. During my daily excursions along the shore this beautiful little bird was always to be met with, sometimes sailing over me uttering its harsh scream of anger, or again holding itself suspended in the air, with head bent down and expectant gaze, watching for the reappearance of some small fish which it had caught sight of. This is the only representative of its family that breeds on the islands, at least the only one which was found, and was the only species taken during July and August.

Breeds abundantly on some of the islands.

101. *STERNA MACROURA, Naum.*
(ARCTIC TERN.)

I did not observe this Tern. Audubon states, however, that when he visited the Magdalen Islands in June, 1833, "some dozens of Arctic Terns were plunging into the waters, capturing a tiny fish or shrimp at every dash."

Mr. C. J. Maynard writes me that he found *Sterna macroura* on the islands, but it was not as numerous as the last species.

102. *PROCELLARIA PELAGICA, L.*
(STORMY PETREL.)

Common, a short distance out at sea. On August 21, just after a heavy storm, I killed a fine specimen from the beach at House Harbor.

103. *OCEANITES OCEANICA, (Kuhl.) Coues.*
(WILSON'S PETREL.)

Observed what I believed to be this species while making the passage to the islands.

104. *COLYMBUS TORQUATUS, Brunn.*
(GREAT NORTHERN DIVER.)

Is rather common. On August 2 I saw a number of them near Wolf Island, and killed two; both were immature birds. Breeds.

105. *UTAMANIA TORDA, (L.) Leach.*
(RAZOR-BILLED AUK)

Breeds in great numbers on Bird Rock, and is not uncommon on Byron Island. The very young birds are covered with a blackish down, having a white belly.

106. *FRATERCULA ARCTICA, (L.) Steph.*
(COMMON PUFFIN.)

Breeds abundantly on Bird Rock and Byron Island. One morning I counted fifty-four of them sitting in a line on one of the ledges. They are quite tame, and I have often approached within a few feet of them before they offered to fly.

107. *MERGULUS ALLE*, (*L.*) *Vieill.*

(Sea Dove.)

Mr. Whalen showed me a specimen which he killed on Bird Rock. He stated that they were occasionally seen in winter.

108. *URIA GRYLLE*, (*L.*) *Brunn.*

(Black Guillemot.)

A very abundant species. I arrived too late to find its eggs, but it breeds commonly at Wolf Island and Cap au Meule. A few breed on Byron Island, but none have been found on Bird Rock.

109. *LOMVIA TROILE*, (*L.*) *Brandt.*

(Common Guillemot.)

Very abundant on Bird Rock. Some few were found on Byron Island, but I did not observe any on the other islands. The young have their upper parts covered with a blackish down, mixed with silver gray; the under parts are white.

APPENDIX.

APPENDIX.

A NUMBER of species which we did not procure undoubtedly visit the islands. A naturalist visiting them during the months of May and June would reap a rich harvest, and many interesting facts might be learned regarding the breeding of some of our rarer species.

I include below a few which certainly would be found at some seasons of the year. Most of them are included from descriptions given me by the islanders.

1. *PLECTROPHANES NIVALIS.*

2. *ASTUR ATRICAPILLUS*, (*Wils*) *Bp.*

3. *STEGANOPUS WILSONII*, (*Sab*) *Cs.*

Procured a specimen at Malpeque, P. E. I , in 1875.

4. *LOBIPES HYPERBOREUS*, (*L.*) *Cuv.*

5. *PHALAROPUS FULICARIUS*, (*L*) *Bp.*

6. *TRINGA MARITIMA, Brunnich.*

Taken on P. E. I. in 1875.

7. *BUCEPHALA CLANGULA, (L) Gr.*

8. *BUCEPHALA ALBEOLA, (L) Bd.*

9. *SOMATERIA SPECTABILIS, (L.) Leach.*

An old gunner, who was familiar with the Eider Duck, stated that he had sometimes killed them "with a yellow bunch on the bill."

10. *ŒDEMIA AMERICANA, Sw.*

11. *ŒDEMIA FUSCA, (L.) Sw.*

I was told that late in the fall a large gray Duck, with partly white wings, was common ; probably the female of this species.

12. *ŒDEMIA PERSPICILLATA, (L.) Fleming.*

13. *GRACULUS DILOPHUS, (Sw.) Gray.*

14. *LARUS GLAUCUS, Brunn.*

15. *LARUS EBURNEUS, Gm.*

The following species, none of which are included in the Catalogue, were taken at Malpeque, P. E. I., in 1875, during the months of August and September. It is possible that some of them may occasionally wander to the Magdalen Islands:—

1. *CYANURUS CRISTATUS.*

2. *PERISOREUS CANADENSIS.*

3. *STEGANOPUS WILSONI.*

4. *TRINGA MARITIMA.*

5. *ARDEA VIRESCENS.*

6. *NYCTIARDEA GRISEA var. NAEVIA.*

7. *PORZANA NOVEBORACENSIS.*

8. *MERGUS MERGANSER.*

9. *MERGUS CUCULLATUS.*

10 *STERNA FORSTERI.*

INDEX.

INDEX.

A.

	PAGE	NO.
Acadian Owl,	54	55
Accipiter fuscus,	55	57
Ægiothrus linaria,	42	27
Ægialitis melodus,	61	64
Ægialitis semipalmatus,	61	63
Ampelis cedrorum,	40	22
Anas obscura,	69	86
Anorthura troglodytes,	36	10
Aquila chrysætus,	57	59
Arctic Tern,	76	101
Ardea herodias,	67	82
Astur atricapillus,	81	2
Auk, Razor-billed,	77	105

B.

	PAGE	NO.
Bank Swallow,	40	21
Barred Owl,	53	51
Bay-winged Bunting,	43	30
Bittern,	68	83
Black Brant,	69	84
Black Duck,	69	86
Black Guillemot,	78	108
Black-and-white Creeper,	37	12
Black-and-yellow Warbler,	38	16
Black-bellied Plover,	57	61
Black-billed Cuckoo,	51	45

	PAGE	NO.
Black-capped Chickadee,	35	45
Black-poll Warbler,	38	15
Blue Heron,	67	82
Blue-winged Teal,	70	88
Bonaparte's Gull,	75	99
Botarus minor,	68	83
Brant Goose,	69	84
Branta bernicla,	69	84
Branta canadensis,	69	85
Bunting, Bay-winged,	43	30

C.

	PAGE	NO.
Calidris arenaria,	65	74
Canada Goose,	69	85
Carpodacus purpureus,	41	24
Catbird,	34	4
Cedar Bird,	40	22
Ceryle alcyon,	50	40
Charadrius virginicus,	58	62
Chickadee,	35	7
Chipping Sparrow,	45	34
Chrysomitris pinus,	42	28
Chat,	34	5
Circus cyaneus,	55	56
Coccyzus erythrophthalmus,	51	45
Colymbus torquatus,	77	104
Common Cormorant,	73	95

	PAGE	NO.
Common Crossbill,	42	26
Common Crow,	48	40
Common Guillemot,	78	109
Common Puffin,	77	106
Common Tern,	75	100
Cormorant,	73	95
Corvus americanus,	48	40
Corvus corax,	47	39
Cotyle riparia,	40	21
Crossbill, Common,	42	26
Crossbill, White-winged,	41	25
Crow,	48	40
Cuckoo, Black-billed,	51	45
Curlew, Esquimaux,	67	81
Curlew, Hudsonian,	67	80
Curlew, Long-billed,	66	79

D.

	PAGE	NO.
Dendrœca æstiva,	37	13
Dendrœca coronata,	38	14
Dendrœca maculosa,	38	16
Dendrœca striata,	38	15
Diver, Great Northern,	77	104
Double-crested Cormorant,	82	12
Downy Woodpecker,	51	47
Dove, Sea,	78	107
Duck Hawk,	56	
Duck, Black,	69	85
Duck, Dusky,	69	85
Duck, Eider,	71	92
Duck, Harlequin,	71	91
Duck, Long-tailed,	71	90
Dunlin,	64	72
Dusky Duck,	69	85

E.

	PAGE	NO.
Eagle, Golden,	57	59
Eider Duck,	71	92
Empidonax flaviventris,	50	43
Empidonax minimus,	49	42
Empidonax traillii,	49	41
English Snipe,	62	66
Eremophila alpestris,	37	11
Ereunetes pusillus,	63	68
Esquimaux Curlew,	67	81

F.

	PAGE	NO.
Falco communis,	56	
Falco columbarius,	56	58
Falcon, Peregrine,	56	
Finch, Purple,	41	24
Flycatcher, Least,	49	42
Flycatcher, Traill's,	49	41
Flycatcher, Yellow-bellied,	50	43
Fox Sparrow,	46	37
Fratercula arctica,	77	106
Fulegula marila,	71	89

G.

	PAGE	NO.
Gallinago Wilsoni,	62	66
Gannet,	72	94
Glaucus Gull,	82	13
Godwit, Hudsonian,	65	75
Golden Eagle,	57	59
Golden Plover,	58	62
Golden-crowned Kinglet,	35	6

	PAGE	NO.
Golden-crowned Thrush,	39	7
Golden-winged Woodpecker,	52	48
Goose, Brant,	69	84
Goose, Canada,	69	85
Goshawk,	81	1
Grackle, Rusty,	47	38
Graculus carbo,	73	95
Great Black-backed Gull,	73	96
Great Blue Heron,	67	82
Great Northern Diver,	77	104
Greater Blackhead,	71	89
Greater Yellowlegs,	65	76
Green-winged Teal,	70	87
Grosbeak, Pine,	41	23
Gull, Bonaparte's,	75	99
Gull, Great Blackback,	73	96
Gull, Kittiwake,	74	98
Gull, Glaucus,	82	13
Gull, Herring,	74	97
Guillemot, Common,	78	109
Guillemot, Black,	78	108

H.

Hairy Woodpecker,	51	46
Heralda glacialis,	71	90
Harlequin Duck,	71	91
Harrier, Marsh,	55	56
Hawk Owl,	53	53
Hawk, Marsh,	55	56
Hawk, Sharp-shinned,	55	57
Hawk, Pigeon,	56	58
Hawk, Duck,	56	
Hermit Thrush,	33	2

	PAGE	NO.
Heron, Great Blue,	67	82
Herring Gull,	74	97
Histrionicus torquatus,	71	91
Hudsonian Curlew,	67	80
Hudsonian Godwit,	65	75
Hudsonian Titmouse,	36	8

I.

Indian Hen,	68	83
Ivory Gull,	82	14

J.

Junco hyemalis,	44	33

K.

King Eider,	82	8
Kingfisher, Belted,	50	44
Kinglet, Golden-crowned,	35	6
Kittiwake Gull,	74	98
Knot,	64	73

L.

Lagopus albus,	57	60
Lark, Shore,	37	11
Larus argentatus,	74	97
Larus marinus,	73	96
Larus philadelphia,	75	99
Larus tridactylus,	74	98
Least Flycatcher,	49	42
Least Sandpiper,	63	69
Limosa hudsonica,	65	75

	PAGE	NO.
Linnet, Red-poll,	42	27
Linnet, Pine,	42	28
Little Auk,	78	107
Lomvia troile,	78	109
Long-billed Curlew,	66	79
Long-eared Owl,	52	49
Long-tailed Duck,	71	90
Loon,	77	104
Loxia curvirostra,	42	26
Loxia leucoptera,	41	25

M.

	PAGE	NO.
Macrorhamphus griseus,	62	67
Marsh Hawk,	55	56
Melospiza palustris,	43	31
Melospiza melodia,	44	32
Merganser, Red-breasted,	72	93
Mergus serrator,	72	93
Mergulus alle,	78	107
Mimus carolinensis,	34	4
Minotilta varia,	37	12
Murre,	78	109

N.

	PAGE	NO.
Nuthatch, Red-bellied,	36	9
Numenius borealis,	67	81
Numenius hudsonicus,	67	80
Numenius longirostris,	66	79
Nyctale acadica,	54	55
Nyctale tengmalmi,	54	54
· Nyctia nivea,	53	52

O.

	PAGE	NO.
Oceanites oceanica,	76	103
Œdemia americana,	82	10
Œdemia fusca,	82	11
Olive-backed Thrush,	34	3
Otus vulgaris,	52	49
Oven Bird,	39	17
Owl, Acadian,	54	55
Owl, Barred,	53	51
Owl, Hawk,	53	53
Owl, Long-eared,	52	50
Owl, Short-eared,	52	49
Owl, Richardson's,	54	54
Owl, Snowy,	53	52

P.

	PAGE	NO.
Parus atricapillus,	35	7
Parus hudsonicus,	36	8
Passerculus savanna,	43	29
Passerella iliaca,	46	37
Pectoral Sandpiper,	64	70
Peregrine Falcon,	56	
Petrel, Stormy,	76	102
Petrel, Wilson's,	76	103
Phalarope, Wilson's,	81	2
Phalarope, Northern,	81	3
Picus pubescens,	51	47
Picus villosus,	51	46
Pigeon Hawk,	56	58
Pine Grosbeak,	41	23
Pine Linnet,	42	28
Pinicola enucleator,	41	23
Piping Plover,	61	64.

	PAGE	NO.
Plover, Black-bellied,	57	61
Plover, Golden,	58	62
Plover, Piping.	61	64
Plover, Ringneck,	61	63
Pooecetes gramineus,	43	30
Procellaria pelagica,	76	102
Ptarmigan, Willow,	57	60
Puffin,	77	106
Purple Finch,	41	24
Purple Sandpiper,	81	5

Q.

Querquedula carolinensis,	70	87
Querquedula discors,	70	88

R.

Raven,	47	39
Razor-billed Auk,	77	105
Red Crossbill,	42	26
Red Phalarope,	81	4
Red-bellied Nuthatch,	36	9
Red-breasted Merganser,	72	93
Red-breasted Sandpiper,	64	73
Red-breasted Snipe,	62	67
Red-poll Linnet,	42	27
Redstart,	39	19
Regulus satrapa,	35	6
Richardson's Owl,	54	54
Ringneck,	61	64
Robin,	33	1
Rusty Grackle,	47	38

S.

	PAGE	NO.
Sanderling,	65	74
Sandpiper, Least,	63	69
Sandpiper, Pectoral,	64	70
Sandpiper, Sanderling,	65	74
Sandpiper, Semipalmated,	63	68
Sandpiper, Spotted,	66	78
Sandpiper, Red-breasted,	64	73
Sandpiper, White-rumpled,	64	71
Savanna Sparrow,	43	29
Saw-whet Owl,	54	55
Saxicola œnanthe,	34	5
Scolecophagus ferrugineus,	47	38
Sea Dove,	78	107
Seiurus aurocapillus,	39	17
Seiurus noveboracensis,	39	18
Sitta canadensis,	36	9
Semipalmated Sandpiper,	63	68
Semipalmated Plover,	61	63
Setophaga ruticilla,	39	19
Sharp-shinned Hawk,	55	57
Shore Lark,	37	11
Short-eared Owl,	52	50
Snipe, Wilson's,	62	66
Snipe, Red-breasted,	62	67
Snowbird,	44	33
Snow Bunting,	82	15
Snowy Owl,	53	52
Somateria mollissima,	71	92
Sparrow, Chipping,	45	34
Sparrow, Fox,	46	37
Sparrow, Savanna,	43	29
Sparrow, Song.	44	32
Sparrow, Swamp,	43	31
Sparrow, White-crowned,	46	36

	PAGE	NO.		PAGE	NO.
Sparrow, White-throated,	45	35	Traill's Flycatcher,	49	41
Spizella socialis,	45	34	Tringa alpina,	64	72
Spotted Sandpiper,	66	78	Tringa bonapartei,	64	71
Squatarola helvetica,	57	61	Tringa canutus,	64	73
Sterna hirundo,	75	100	Tringa maculata,	64	70
Sterna macroura,	76	101	Tringa minutilla,	63	69
Stone Chat,	34	5	Tringa maritima,	81	6
Stormy Petrel,	76	102	Tringoides macularius,	66	78
Strepsilas interpres,	62	65	Turdus migratorius,	33	1
Sula bassana,	72	94	Turdus pallasi,	33	2
Summer Warbler,	37	13	Turdus swainsoni,	34	3
Surnia ulula,	55	53	Turnstone,	62	65
Swallow, Bank,	40	21			
Swallow, White-bellied,	40	20			
Swamp Sparrow,	43	31	U.		
Syrnium nebulosum,*	53	51			
			Uria grylle,	78	108
			Utamania torda,	77	105

T.

W.

	PAGE	NO.		PAGE	NO.
Tachyceneta bicolor,	40	20	Warbler, Black-poll,	38	15
Teal, Blue-winged,	70	88	Warbler, Black-and-yellow,	38	16
Teal, Green-winged,	70	87	Warbler, Summer,	37	13
Tern, Arctic,	76	101	Warbler, Yellow-rumpled,	38	14
Tern, Common,	75	100	Water Thrush,	39	18
Thrush, Golden-crowned,	39	17	Wheatear.	34	5
Thrush, Hermit,	33	2	White-bellied Swallow,	40	20
Thrush, Olive-backed,	34	3	White-crowned Sparrow,	46	36
Thrush, Water,	39	18	White-rumpled Sandpiper,	64	71
Titmouse, Hudsonian,	36	8	White-throated Sparrow,	45	35
Totanus flavipes,	66	77	White-winged Crossbill,	41	25
Totanus melanoleucus,	65	76	Wild Goose,	69	85

* For *Syrnuim nebulosum*, page 53, read *Syrnium nebulosum*.

	PAGE	NO.
Willow Ptarmigan,	57	60
Wilson's Petrel,	76	103
Wilson's Phalarope,	81	3
Wilson's Snipe,	62	66
Wilson's Tern,	75	100
Winter Wren,	36	10
Woodpecker, Downy,	51	47
Woodpecker, Golden-winged,	52	48
Woodpecker, Hairy,	51	46
Wren, Winter,	36	10

Y.

	PAGE	NO.
Yellow Warbler,	37	13
Yellow-bellied Flycatcher,	50	43
Yellow-rumpled Warbler,	38	14
Yellow-shanks,	66	77

Z.

	PAGE	NO.
Zonotrichia albicollis,	45	35
Zonotrichia leucophrys,	46	36

THE END.

www.ingramcontent.com/pod-product-compliance
Lightning Source LLC
Chambersburg PA
CBHW021947190326
41519CB00009B/1174